HOW TO CREATE A GREAT INFORMATION TECHNOLOGY VISION --
And Thrill Your CEO

THOMAS S. IRELAND

All rights reserved. No part of this book may be reproduced or transmitted in any form by any means without written permission of the author.

Copyright © 2012 Thomas S. Ireland

DEDICATION

To my children, Heather, Tara, Brendan and Ryan

Contents

Introduction 1
  Who is This Book For? 1
Why I Feel Compelled To Write This Book 3
  A Note About Leadership 4
  Leadership and Ethics 6
  Housekeeping 7
In the Beginning 8
  My Definitions 8
  The Hierarchy 8
The Elements of Mission 10
The Elements of Vision 13
  The Relationship of Business and Technical Vision 18
  Enterprise Mission Is The Platform 18
  The Primary Driver Is The Enterprise Strategic Vision . 19
  The Feedback Process Shapes The Vision 27
The Necessity for Integrated Vision 35
  The Danger Area 35
  Danger Area Breakout 36
  Breakout and Feedback 40
  Planning Frequency 43
  The Optimum Vision Development Path 44
  The Distance To Our Goal 46
Business First; Technology Second 51
Leadership Before Technology 53
The One Page Guiding Document 57
About The Author 63

ACKNOWLEDGEMENTS

With great thanks to John Langenbahn from whom I borrowed so much in writing this book. He is the greatest leader I have ever personally met or worked for. I am honored that he calls me his friend.

Introduction

Who is This Book For?

This book is written for the information technology, or IT, professional at the manager level and above. However it can be very useful to the person who is not in the information technology profession.

Every manager, director, vice-president and C-level professional is required at several times during their career to write a Vision statement or a Mission statement. The first time a person is given this task, it is almost certainly a confusing effort. The first questions you might have are, "What is a Vision statement?" followed quickly by, "What is a Mission statement?" and "What is the difference between them?"

Once you begin to understand the answers to these questions the real difficulty begins. The statement developer must write something that supports enterprise objectives established by the chief executive and their leadership team. At the same time, the writer wants to be sure that the statement and supporting documentation they put together stimulate their employees to action. Employees will see right through Mission and Vision statements that are just words written by their manager to complete an unwanted assignment. However, they will be motivated by, even self-motivated by, well-developed, meaningful documents developed by leaders who believe strongly in what they have written.

Information technology is an extremely fast-paced profession. In recent decades, IT has become increasingly recognized as an important and critical major business tool within enterprises. It is this recognition that resulted in the creation of the Chief Information Officer position. At the same time, each member of an information technology must understand, and fully internalize, something that may sound like heresy. When writing an IT Mission or Vision statement in support of an Enterprise the writer must constantly keep in mind that technology without business application has no value!

In most cases, information technology is evolving much more quickly than the organization in which the IT team is embedded. Because IT is fast moving, it can be used as an accelerator for the entire enterprise and perhaps be used to improve the enterprise's position in its industry.

In this book, I'm going to show you how to develop Mission and Vision statements that do much more than just describe how your IT team can support your enterprise. I'm going to show you how you can define the means by which your team can be a critical tool to help accelerate your enterprise to greatness!

Why I Feel Compelled To Write This Book

After ten years working for the Air Force as an officer and civilian I went to work for a very demanding information technology company. We were running hard and fast. When we were relatively small we all communicated well and were heading in the same direction. Company growth was about 40% per year. The network group I managed was growing at 100% per year. If we were going to keep working as a cohesive unit as we grew we needed to go through the pain of developing greater formality in structure and direction. That is a tough transition for a small entrepreneurial company.

That painful day came when my boss, in a significant amount of confusion and consternation, came to me and said that he and I were going to work together on Vision and Mission statements for his organization. We didn't even know where to begin! To make things worse, the senior executives were just beginning to write their Vision and Mission statements! We didn't know how to do what we were asked and we didn't know what to do because we had no platform to work from. Finally, all of us at all levels in the organization finally just muddled and turned in stuff just to get the impossible project off our plates. Needless to say, our Mission and Vision statements weren't very motivational and added nothing of value to the organization.

However, a seed was planted within me. In spite of very complex and demanding missions in the Air Force I always had a good idea where I was headed. The core of the Air Force Mission statement was "to fly, fight, and win". The Vision statement was about one paragraph long but just as clear. Every organization had a Vision and a Mission statement. Each one supported the next higher organization in a very logical and coherent manner. I knew as an information technology Air Force officer exactly what my teams had to support and what our objectives were. Contrary to popular belief, we had a lot of liberty and were expected to take a lot of initiative to achieve our Mission and realize our Vision. Our objectives were always clear and we were well-focused even when the situation periodically became more tense.

I recognized, when that seed was planted, that our information technology company, and my team, badly needed a clear, strong Vision and Mission if we were going to be motivated, efficient and effective. So, I made sure from that day onward that my team always knew where we were going strategically and tactically. We operated with a motivational, realistic Vision and Mission. The results were extremely positive even though the people "upstairs" never provided the same formal statements for the overall enterprise.

A Note About Leadership

We all have a pretty good idea of where executives want us to go even if they don't always define it as well as they should. Leadership can be practiced at all levels of an organization. If you tell people clearly what the objective is they will diligently work to reach that objective. Naturally, within the context of this book, clear Mission and Vision statements help to define that objective.

Behind all of this, however, is one thing of extreme importance. Any leader at any level in the organization must have an expectation of success for the people or department they manage. I believe that the first rule of leadership is "People will meet your expectations." This is true whether those expectations are high, low or somewhere in between. That expectation of success must be felt strongly in the Mission and Vision statements and in all rhetoric and action associated with those statements.

Leadership is necessary at all levels. A manager working for a weak or inexperienced director can't wait for that person to lead if it is clear that nothing in this world can make them do so. The leaderless manager must fill the gap and do everything they can to provide leadership to the group they are responsible for.

Technology groups seem to have a habit of promoting great technologists into leadership positions as a reward for having accomplished excellent technical work. This is an epidemic within the IT industry. As a result, a person who has excellent leadership ability might find themselves working for a superior technologist who doesn't have a clue about leadership. That is just the way it is in the IT business. The person with leadership talent who finds that they are working for a non-leader simply has to exert leadership within their own department without the kind of guidance and mentoring they deserve and desire.

You will sometimes have to write Mission and Vision statements without an appropriate template from your boss or even higher up the organization. However, if you put these important documents together in an honest manner for the team you manage they will meet your high expectations and you will add great value to your enterprise.

Leadership and Ethics

Every business has a set of ethics. Some executives state their ethical standards more clearly than others and the standards may vary officially or unofficially. I have found that businesses with the highest standards are the most pleasant to work for and achieve the greatest level of success.

The clearest standard I've encountered, and which the corporation lived up to was "It is only be dealing honestly and fairly in all things that real success is attainable." That quote is from George Mead and was posted clearly in the main lobby of the Mead Paper Company in Dayton, Ohio. It set a standard that meant a lot to me and which most of us worked hard to live up to. As a consequence, business people in other companies knew that we would only work toward equitable and ethical business deals. It was part of our culture at all levels and it penetrated everything we did. It even influenced the quality of our products and how we treated each other.

In another example, I once worked for a company that made quality delivery of product to the end customer its highest priority. The customer was the most important person in the organization. The customer rewarded us with their loyalty and the business prospered. While I was there we were acquired by a larger organization with a different set of priorities. Almost overnight our focus was forced to shift from the customer to assuring optimum financial compensation for our executives. Over the next two years we lost market share and the business failed.

Housekeeping

Please note that I've deliberately capitalized the first letter of each key word in this book, such as Mission and Vision to emphasize its importance in the planning process.

In the Beginning

My Definitions

You can find many definitions of Mission and Vision. They are all good definitions within the context of how they are being used in various books. The definitions I will use in this book form a platform that I intend to be simple and effective for information technology teams. They are also a platform for, and consistent with, other books written by this author for information technologists

Mission defines an organization. It states why the organization exists. When an organization is in its infancy, or perhaps is still an unfulfilled idea in the mind of an entrepreneur, Mission and Vision may be intertwined. But eventually the description of what the business will deliver will evolve out of the initial inspiration. In this book we are going to focus on the business description as being the fulfillment of the idea. Beyond that there has to be ongoing inspiration to move the business forward. That ongoing inspiration is a periodically renewed Vision.

The Hierarchy

Once the fledgling organization begins to take better form Mission and Vision finally begin to separate from each other.

When the entrepreneur begins developing an idea for a new enterprise everything about that idea is just a bit ball of concept. The entrepreneur understands only in general terms what they want to do and may only have a glimmering about how to get it done. Mission, Vision and everything else about the enterprise is still enmeshed in this vague ball of concept. As the concept evolves, Mission and Vision begin to develop their own distinct characteristics. The entrepreneur can define in more concrete and specific terms why the organization is going to exist and can explain that for potential investors in the form of a written Mission. Out of that Mission statement some substance can be created in very general and high level terms. That substance becomes the initial Vision that describes how the Mission will be accomplished. We will see in this book that Vision can be tactical or strategic. At this early point in the evolution of an enterprise Vision is, of necessity, strategic.

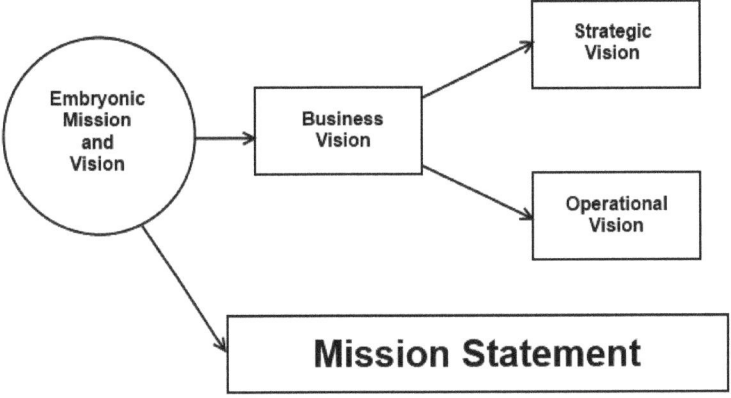

The Elements of Mission

As mentioned earlier, there are many different definitions of Mission and Vision. And admittedly, distinguishing between Mission and Vision can be difficult. For purposes of this book, and other books by this author for information technologist, I'm offering definitions for Mission and Vision that are very distinct from each other. I've worked to keep the definitions of Mission and Vision in this book simple and practical. These definitions are based upon forty years of experience and success in their application across a number of enterprises.

Mission is the reason an enterprise exists and is very stable and very clear over time. The mission of the United States Air Force is "to fly, fight and win". That has been consistent over a long period of time. The Vision of how to accomplish that Mission has evolved over time as technology and threats changed.

In the context of this book, and other related books by the author, a planning hierarchy is clearly established. A Mission has to be defined before a Vision can be created. At that point the Vision becomes the stable planning platform unless the enterprise goes through a Mission change. Mission changes are very rare! Vision is adjusted at every planning cycle. During the life of that planning cycle the Infrastructure, Tactical Path and Strategic Target become defined and are all based on that Vision.

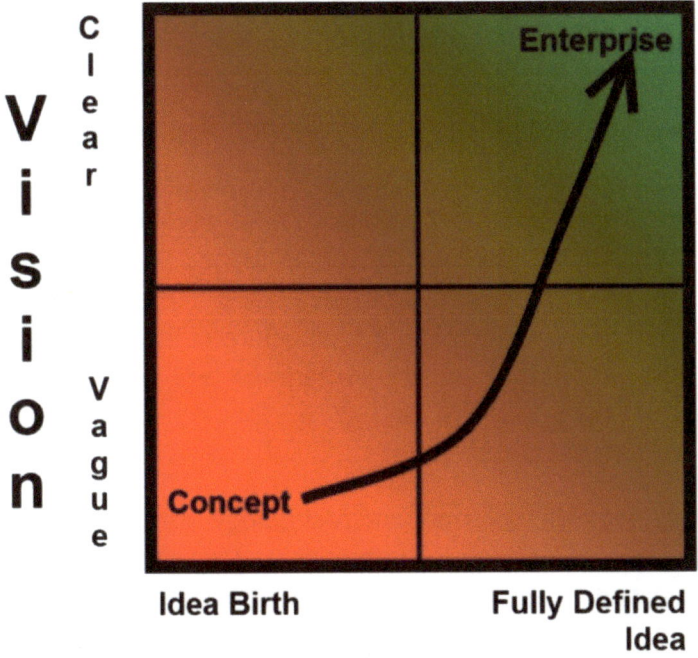

Mission

In very compact terms the Mission is <u>WHAT</u> an enterprise is today and <u>WHY</u> it exists. It is a very stable statement in that it doesn't change significantly over the near term. It might remain the same through several Strategic Planning cycles. The Vision is <u>HOW</u> the Mission is going to be accomplished and may change with each planning cycle.

In the very early stages of a developing enterprise the idea, which is the Mission, must become better defined before a good Vision statement can be written.

The Mission Statement of the enterprise is long lasting. The Business Vision and the Business Mission Statement are both born out of the conceptual ball that is the Embryonic Mission and Vision. This is the proverbial idea written on a napkin in an empty cafeteria in Las Vegas at two o'clock in the morning. Once the enterprise begins to take better form, the Business Vision begins to separate itself into Strategic and Operational Visions.

Once an entrepreneur's concept becomes an enterprise the Vision becomes the platform for all planning. This is important to the reader because, in this book, everything from the next section onward evolves out of the Vision. The business and technical Infrastructure of the enterprise, and the Tactical Path the enterprise will follow to a Strategic Target at the end of a planning horizon, are based on the Vision platform.

The task of the Information Technology executive is to establish an IT Mission Statement and an IT Vision that support and preferably enhance the Business Mission and Vision.

In summary, The Mission Statement is:

- The description of WHAT and enterprise is today and why it exists,
- A stable statement that doesn't change significantly over the near term,
- Very possibly stable through several Strategic Planning periods,
- The well-defined platform for Vision Statements.

The Elements of Vision

The information technologist who is asked to create a Vision for their department or organization will want to know what their Vision is supporting. In a healthy, well-run enterprise that technology manager will be given a Mission Statement and a Vision Statement for the enterprise and all elements of the organization above their level. Unhappily, about 50% of the time this information won't be available. Sometimes this is because somebody "upstairs" will have decided for some reason that everyone must have Vision statements where none have existed before. Every manager at all levels might be tasked to do this at the same time. This is an unusual and worst case but it happens.

As information technologists we are used to worst case planning. After all, our systems that we design and operate are required to be survivable. In today's world when the IT infrastructure dies the business is immediately and dramatically impaired. Building this survivability into IT systems requires innovative thought. We can apply that innovative talent in this potential worst case for writing Mission and Vision statements. If we aren't faced with the worst case then the process and methods described here will still work.

Enterprise Vision ranges from Operational through Strategic.

Operational Vision is what is needed to get through the day-to-day, or short range, tasks of producing product or performing the background actions that support production activities.

Strategic Vision is that clearly defined view of the future that states where the business will be at the end of the planning horizon. In this book we will be focused on Strategic Vision. However, it is important to acknowledge and include in the planning process the manager who, of necessity, must work with a day-to-day or Operational Vision.

The operational manager is critical to the business. Some who are reading this book may be an operational manager or have held an operations manager position at one time or another. This is the team player who keeps the daily processes running smoothly, ensures that the raw materials (literally and figuratively) arrive on time, get used properly and are turned into a product that goes out the door on time with high quality. This is not the person who is responsible for the long term Vision of the enterprise. However, this is the person who has years of experience that we draw on when determining how to do new things without disrupting production. It could also be a person, depending on the type of company and operational requirements, who will over time move to a position in which they are eventually responsible for the Vision process in a more strategic manner. The person with Operational Vision is a key individual who must be allowed and encouraged to play their operational position on the team and be consulted about how the Strategic Vision affects the ability of their team to do their job.

This book is written for the person who is, or who wants to be, the visionary leader who is depending on others to do their jobs while they perform their task of understanding best where to take the organization. This is the highly paid person who, if they do their job right, makes it possible for the rest of us to keep our jobs. Their ability to clearly envision the future of the marketplace, the competition and the enterprise means the difference between long term success or failure. Their skill is vital and their decisions may be of broader consequence than those of most others in the organization. Their ability to put all this information into a concise, clear, energizing, action-oriented Vision Statement can make the difference between organizational success or failure over the long term.

In the process described in this, and related books, I'm defining four elements of Vision.

- Business Vision is the long range objective that the top enterprise executives have established for the entire enterprise. Long range in this case means an average of three years and can be up to about five years. This Business Vision sets the stage for Business Infrastructure, Technical Vision and Technical Infrastructure.
- Technical Vision is developed by the chief IT executive in cooperation with the enterprise executive team. It is based on, supports, and may enhance the Business Vision.
- Strategic Vision is that clearly defined view of the future that states where the business will be at the end of the planning horizon. In this book we will be focused on Strategic Vision. However, it is important to acknowledge and include in the planning process the manager who, of necessity, must work with a day-to-day or Operational Vision.
- Operational Vision is what is needed to get through the day-to-day, or short range, tasks of producing product or performing the background actions that support production activities.

As stated above, and in the context of this book, Business Vision is about three to five years into the future. At least it is supposed to be. Hopefully your executive level business leaders will not be focused on day-to-day or operational issues. For your business to be successful your top level executive team must have a Strategic Vision.

That seems obvious but we have probably all, in both the private and public sector, encountered a surprising number of executives who operate from a task list instead of a long term Vision. I once worked for a chief executive who kept a list of tasks that he reported on to our board each month. Each of these tasks was a short term accomplishment designed to show how much work he was doing. The rest of us worked with him and for him on those tasks. He never produced a Strategic Vision although he saw his task list as Strategic. The organization, of course, had not progressed at all while he was there. We accomplished operational tasks but we didn't move forward. Happily he didn't stay with the enterprise very long.

Business Vision, because it is the first and direct offspring of the core Mission, is the basis for all planning in the entire enterprise. In a great organization there will be a clear top level Business Vision on which all planning is based. But, we have already admitted that this ideal situation doesn't exist a significant part of the time. As this book continues we will talk more about the creativity that is required to overcome this impediment to success. In short, what a manager may sometimes have to do is make their best educated determination of what the long range Business Vision is when one isn't supplied. I've had to do this more than once, so it is possible to do this and still achieve a reasonable result.

The focus of this book is on guiding the information technology leader in creating a Technical Vision. The Technical Vision is based on the Business Vision. In terms that the IT readers of this book will appreciate, the Business Vision is the independent variable and the Technical Vision is the dependent variable. As such, they can be represented on the following graph.

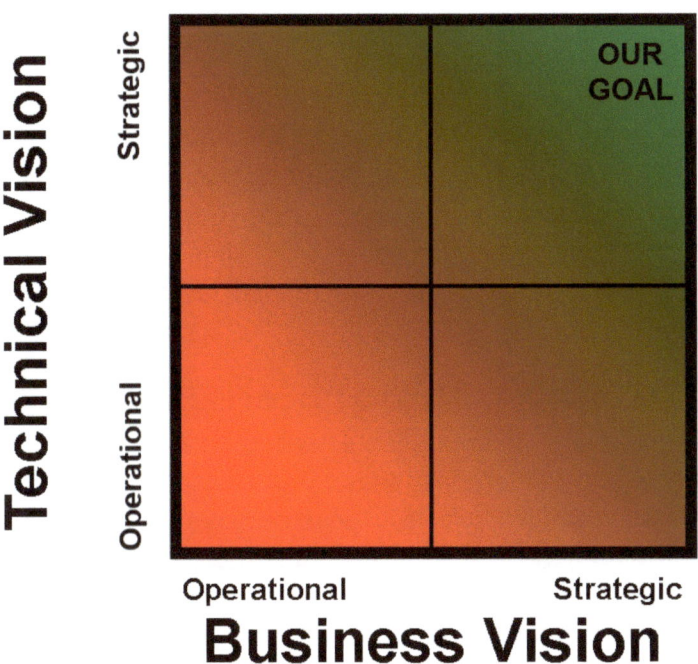

You will notice on both axes of this graph that they run from Operational Vision to Strategic Vision. At the highest levels of a well-run organization there will be an Integrated Strategic Vision provided by both the enterprise and the technology leaders. All of this is against the backdrop of a Mission Statement.

The Relationship of Business and Technical Vision

The method offered in this book is one that consistently and firmly focuses the information technologist's attention on support of the Enterprise Mission and Vision. Information technology is critical, even essential, to the success of almost every modern enterprise. However, it is important to keep in mind that "Technology without business application has no value." That is why throughout this entire book we will keep referring back to Enterprise Mission and Vision as the platform for developing the IT Mission and Vision. In that context, the elements of the tool we will be working through are related to each other as shown in the following figures .

Enterprise Mission Is The Platform

The Enterprise Mission Statement is the platform for everything an enterprise does. It doesn't matter if that enterprise is a private, government or non-profit organization. Our goal is to reach the upper right corner of the Vision quadrant. If we don't know the Mission then we don't know <u>What</u> we are supposed to do. If we don't know <u>What</u> we are supposed to do then we can't possibly know <u>How</u> to do it! In the context of this book Vision is one of the key elements that define <u>How</u> we are going to accomplish the Mission. This part of our discussion is going to focus on the four elements of Vision to drive toward what the IT team has to do to support the Enterprise Mission

The Primary Driver Is The Enterprise Strategic Vision

We start with the Enterprise Strategic Vision because that is the first workable piece that falls out of the original conceptual ball written on that napkin we mentioned earlier.

There are clearly a number of high level elements that come together when creating the Enterprise Strategic Vision. The CEO and the executive team have to make decisions on a number of things that give some practical definition to the Business Vision. A few of these, and by no means all, are shown in the following chart.

All of these elements must be addressed in the long view and they don't have to be set in concrete. Marketplaces change quickly for a wide variety of reasons. Some of those reasons are predictable. Sometimes change will happen for reasons that no one has a ghost of a chance or predicting. The longer range the prediction the more volatile the deviation from that prediction might be. The key is choosing a time frame that is far enough into the future to be considered Strategic but not so far as to fall into the realm of ludicrous guesswork. As the pace of change increases the number of years involved in predictive planning is becoming less. A plan that was seven or eight years long in the 1980's might have had fairly accurate results. Today, such a lengthy prediction might readily take an enterprise down a horribly wrong path.

> **Enterprise Strategic Vision**

> Market Segment
> Strategic Partnerships
> Geographic Reach
> Business Ethic
> Products and Product Integration
> Financial Targets

The market has to be narrowed down to some segment of seven continents and six billion people. Financial goals have to be realistically established that support the business's ability to competitively deliver a quality product to market. A business ethic has to be defined that employees can be proud of and adheres to the laws and customs of the countries in which the company is selling and setting up offices. Vendor partnerships might have to be created that optimize cash flow and productivity. There may be considerations in a very large corporation of internal product relationships as well as external product relationships. The list is much longer than the few examples given here.

However, when that list is well thought out and established the executive team will have the opportunity to write out and explain a clear Enterprise Vision Statement to the rest of the organization. In this initial iteration, this is a draft working document which will be used by middle management to create supporting Operational Visions. In the core enterprise functions this will be called the Enterprise Operational Vision.

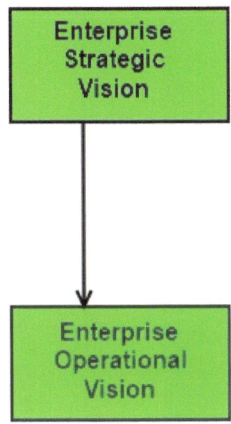

Market Tactics and Focus
Tactical Partnerships
Pattern of Geographical Expansion
Corporate Culture
Product Development and Evolution
Income Diversification

Each element of the Enterprise Operational Vision must reflect and support corresponding elements of the Enterprise Strategic Vision. The enterprise middle management must determine the tactics they will use to develop the market segments the executive staff has targeted. Specific vendors and potential partners must be evaluated and relationships established that support the types of strategic partnerships executive management has determined are necessary for long term success. While the Strategic Vision may define the eventual geographic reach, the people responsible for executing that Vision must develop the best sequence for achieving that goal.

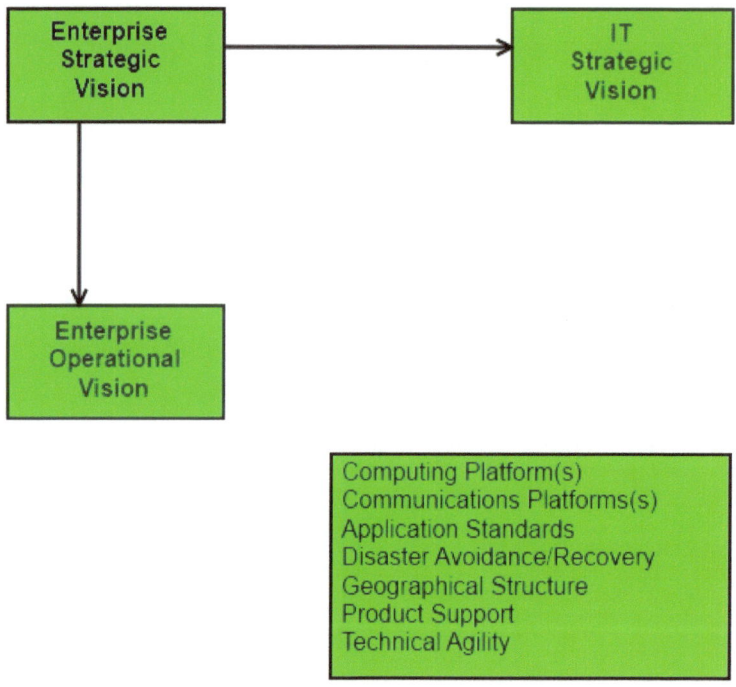

The process described in this book works best in an enterprise that has a clearly documented Enterprise Strategic Vision. We all know that isn't always true. Not every enterprise has leadership that understands the importance of top executive level Mission and Vision statements. But, the IT team must not use that as an excuse to avoid this process.

At the very least there is an Enterprise Vision that can be estimated by the technology team even if it isn't written and documented by the enterprise. It may not be exactly right but it can be close enough to serve as an excellent starting point. Any future revisions can be made during the next planning cycle several years downstream. It may very well be that an IT team that shows such initiative may be the catalyst for the executive team to do their duty.

The IT Executive Management has some tasks in Strategic Vision development that are similar to those of the core enterprise middle management. They must come up with ways to support the Enterprise Strategic Vision. However, IT Executive Management has another responsibility which may be even more important and is unique to the IT team at all levels.

This responsibility is the theme of this book.

In short, the responsibility of the Information Technology team is to not only support the Enterprise Strategic Vision. Its responsibility is to enhance that Vision thus giving the enterprise a competitive advantage. This is where you have the opportunity to thrill your CEO!

The majority of businesses in the world are dependent upon IT for the success of their enterprise. Information Technology is in the back office, the front office and everywhere in between. It connects business partners, customers and vendors. Information Technology is used to track the flow of goods, money, vehicles and even people. It is imbedded in financial, personnel, manufacturing, safety, security, and almost any other application that is a major or minor part of an enterprise. Competition in the private sector is such that proper and aggressive use of IT can make the difference between success and failure or at least the degree of profitability. In local, state and federal government agencies IT can reduce cost of operation thus reducing the need for tax revenue. Perhaps it can even be used to create a better environment for citizens.

The IT executives and managers who develop the supporting IT Strategic and Operational Vision have therefore a different response responsibility than the core enterprise team. Unless the core enterprise is an Information Technology organization it is probable that the enterprise executive team has no understanding of information technology. However, they know that they are vitally dependent upon it. More to the point, an excellent IT staff will know that they can provide technology that can help the enterprise be more successful than perhaps even envisioned by the most aggressive enterprise executive staff.

Things the IT executive staff might consider are the type or types of computing and communications platforms necessary to meet Vision objectives. Application standards must be selected that create a coherent platform across the enterprise to support the selected products in the chosen markets. There must also be enough technical flexibility to meet the evolving mission.

The IT chief executive must be enmeshed in the core enterprise. When they are, they have an increased opportunity to come up with creative ideas to enhance the profitability and competitiveness of the organization.

An example might be the CIO who understands the mainstream processes of an enterprise from converting raw material into product to final distribution of that product to the customer. Such a CIO can overlay their knowledge of information technology onto the business process to potentially eliminate costs and save time.

One real life example is the CIO who determined that a Fortune 150 company could eliminate most warehouses with improved product tracking combined with predictive customer demand. That innovation reduced cost of operation and resulted in improved margins as well as a reduction in price which improved market share.

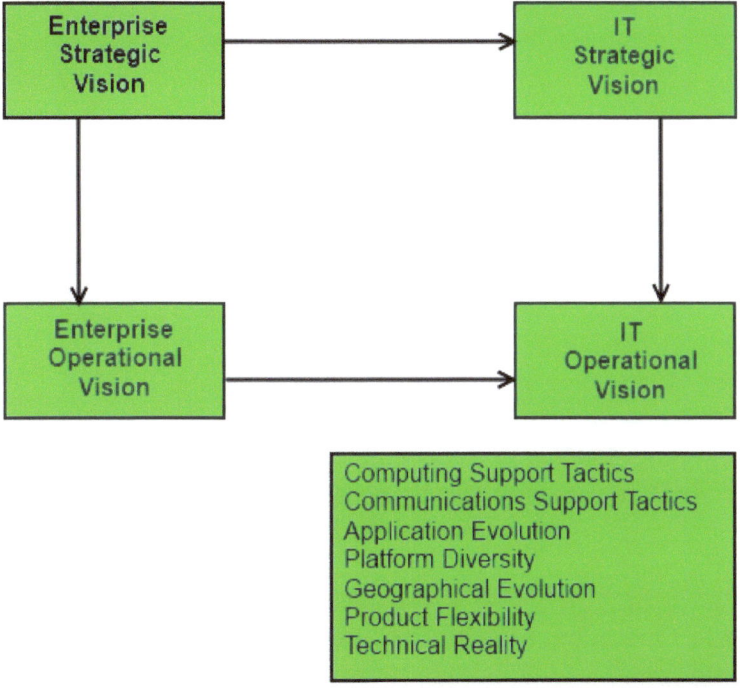

The operational managers on the IT staff must support the IT Strategic Vision as well as respond to the Enterprise Operational Vision. They must determine the details of the computing and communications platforms so that they can support application evolution and application diversity.

Support issues across an evolving geographical and product market must be well thought out and determined. Finally, the realities of technology must be applied against the wishes of the enterprise customers to assure that expectations are properly managed.

Technological reality is a key phrase. Just like anyone else, information technologists can get trapped in an almost religious conviction in support of a particular vendor, applications suite or hardware platform. When this happens they make the technology a higher priority than the business requirement. This can result in higher cost of operation and reduced support to the mission. It is important for each information technologist to understand that **technology without business application has no value!**

As the IT Operational Vision is developed, the tactics employed must reflect the needs of the enterprise Mission and Vision as well as the IT Strategic Vision. The technology platforms used must be able to evolve with the business. The IT operational mission is tough. It requires managing what is perhaps the most rapidly evolving and complex industry to support an enterprise fully dependent upon IT to succeed. Good IT decisions can enhance the success of the enterprise. Bad IT decisions can derail it.

This is why there are arrows in the previous chart pointing at IT Operational Vision from both IT Strategic Vision and Enterprise Operational Vision. In the day-to-day operational mission this group is supporting two masters: their core enterprise customer, and those driving the development of information technology. They also serve both of these masters in the planning process.

The Feedback Process Shapes The Vision

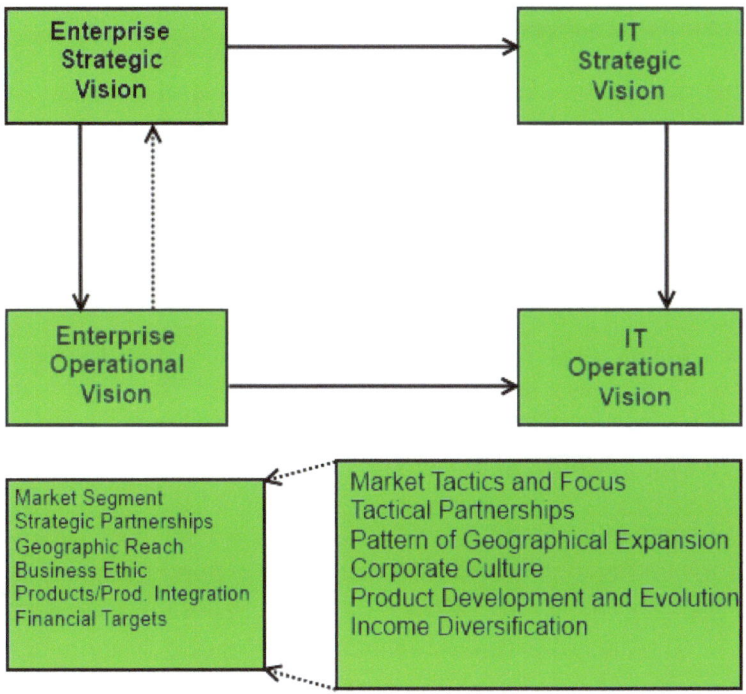

The chief executives can't stand alone in shaping the Enterprise Strategic Vision. Most executives know this and they will seek input from the rest of the team. Traditionally, they will first seek input from their core products operational managers. This first step in the Vision feedback process can provide some adjustment to the Strategic Vision. Strong operational managers will develop tactics and operational methods that make the Strategic Vision possible. More importantly, they will make suggestions that can lead to improvements in the Enterprise Strategic Vision. These improvements can be in the form of reality checks on the plans of the executive leadership which may temper the timing or the focus of the

Vision. Future year financial plans may be tempered by the reality of taking product to market or the ability to generate income on a product line that may need more time to ramp up to needed levels.

This process is fairly routine in enterprises and is critical to shaping a realistic set of plans and expectations. It isn't quite as common to include the Information Technology team this deeply into the process.

It wasn't long ago that excellent CIOs had to fight hard for inclusion in the planning process. Information technology was considered an overhead expense that CEOs often saw as a necessary evil on the bottom line. The CIO's job was regarded as one of cost containment while making sure that systems were running properly.

The best and most visionary CIOs knew that they had a larger responsibility and worked hard to embed themselves in the direct mission of the enterprise. These CIOs were business people first and technologists second. They might or might not have a strong technology background. They may have been introduced into the CIO position from one of the core enterprise divisions. Whatever their background, these CIOs had a Vision that information technology was not just a support tool. They knew it was a business tool that could be used to enhance the enterprise and make it more competitive in the marketplace.

These CIOs and their teams who have integrated themselves fully into the enterprise have been a force for making their organization more successful whether it is a public, private or non-profit agency.

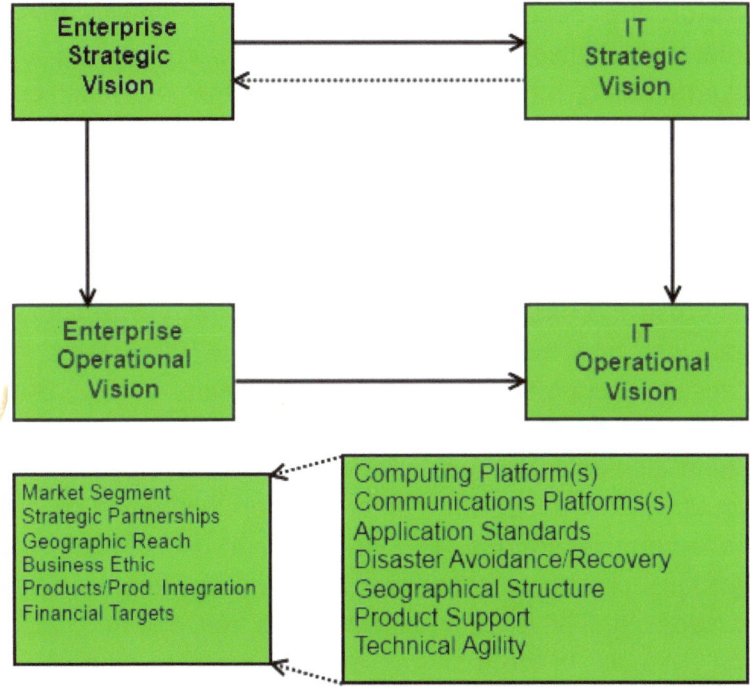

It takes a great deal of insight and talent to fully understand a business requirement at the executive level and strategically apply information technology to support that requirement. However, the objective of this book is to help the technology team do more than just support the enterprise. The objective is to encourage the information technologists to apply their skills in a manner that enhances the enterprise.

The perspective in this book is that the job of an enterprise is to provide the best possible product to a customer at the lowest cost of operation in an ethical and legal manner. The job of the IT team is to help make that happen.

When the CIOs executive team reviews the Enterprise Strategic Vision they will be analyzing how they can provide the necessary technology tools to make their enterprise better than the competition. In addition to understanding the equipment and systems that must be applied to the Enterprise Strategic Vision, the CIOs team must give business and technology feedback that shows how that Vision might be enhanced through the IT tools that can be made available to the enterprise.

Let's use the example in the previous sketch. The CIO's team can take the easy way out and just put systems in place to support the Enterprise Strategic Vision. But wouldn't it be better if the CIO and staff analyzed the Vision elements and carefully chose the technologies that would not just support a market segment but would make the enterprise stronger in that segment? This might be through proactive recommendations on point-of-sale practices, or field sales and inside sales front end and back end systems. In too many enterprises the core enterprise team will approach the IT team with requests to support their need which has already been focused on a particular market. This puts the IT team in a reactive mode limiting the opportunity to create a standard and broad infrastructure on which a multitude of applications can be layered. The proactive CIO and team will involve themselves in the Vision process so that a strong infrastructure can be established that will support the targeted markets with an integrated set of applications that enhance market penetration throughout the entire planning cycle. This approach will create the opportunity to improve margins and market share rather than just provide passive support.

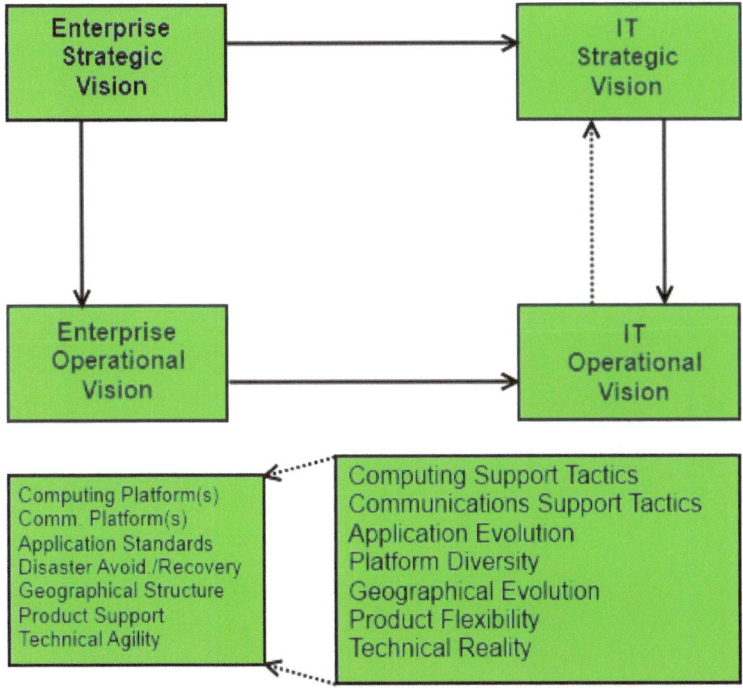

In like manner, the IT middle managers must be echoing the actions of their senior executives. The operational IT tools can't just be developed to assure high quality continuity of operations on a day-to-day basis. Certainly that part is essential. Without that there is no operation. So the detailed elements of the IT Operational Vision must meet those traditional requirements and provide support to the IT Strategic Vision. Once again, a strong, confident IT team will provide feedback that will probably cause adjustments and enhancements to the IT Strategic Vision. In turn this might provide opportunities for improvements to the Enterprise Strategic Vision.

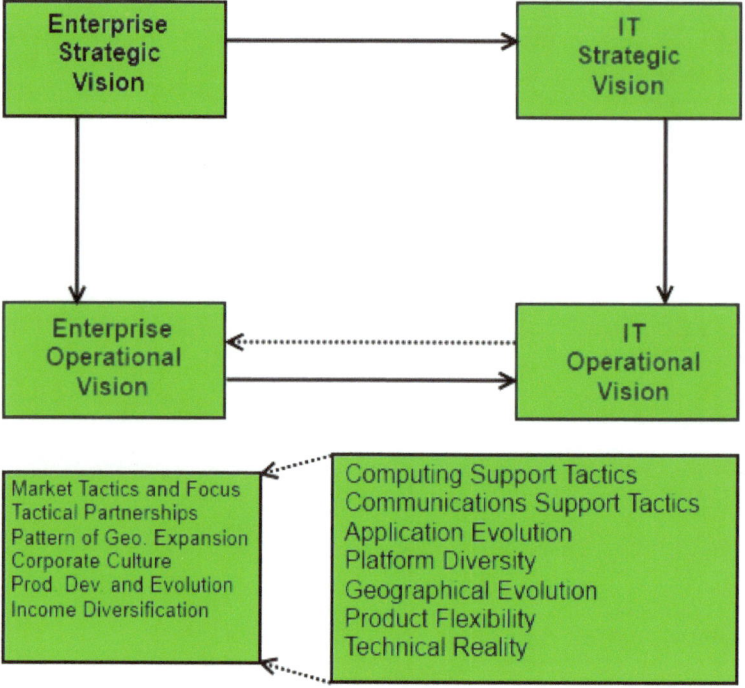

The IT operations team has the opportunity to look across, as well as up, the organization. The IT mid-level team that gets fully involved in the core enterprise operations gives the organization the opportunity to greatly improve efficiency at the production level. An open exchange of information between the product teams and the IT providers will almost certainly lead to new ideas to reduce cost, improve customer service and even provide a more desirable product. Conversely, the IT team may decide to modify to a greater or lesser degree the technology platform they originally intended to use before closely collaborating with the core product people. It is this feedback activity which can give the enterprise a competitive advantage it would not otherwise have.

This type of relationship established when collaborating to optimize the various enterprise Vision statements can be carried through an entire set of activities eventually leading to a more integrated daily operation. Collaborative Vision development sets the stage for a more integrated core enterprise/IT infrastructure. The entire organization can examine this integrated infrastructure during planning cycles to assure that it evolves along a tactical path that limits excess expenditures or dead ends while assuring an optimized competitive platform. All of this becomes the platform for a strategic plan that has a well-developed business and technology foundation.

An example in this area might be the introduction of new technology to provide an outside sales person accurate knowledge of where a product a customer needs is geographically located in the supply chain. This permits the sales person the ability to immediately answer a customer's question about when the product can be delivered. This can mean the difference between closing a sale or giving a sale up to the competition.

Or, as many organizations have done, integrate with a customer's inventory system to automatically ship product when the customer's supplies drop to a certain level. This reduction in customer operational costs can assure a steady stream of revenue for the supplier.

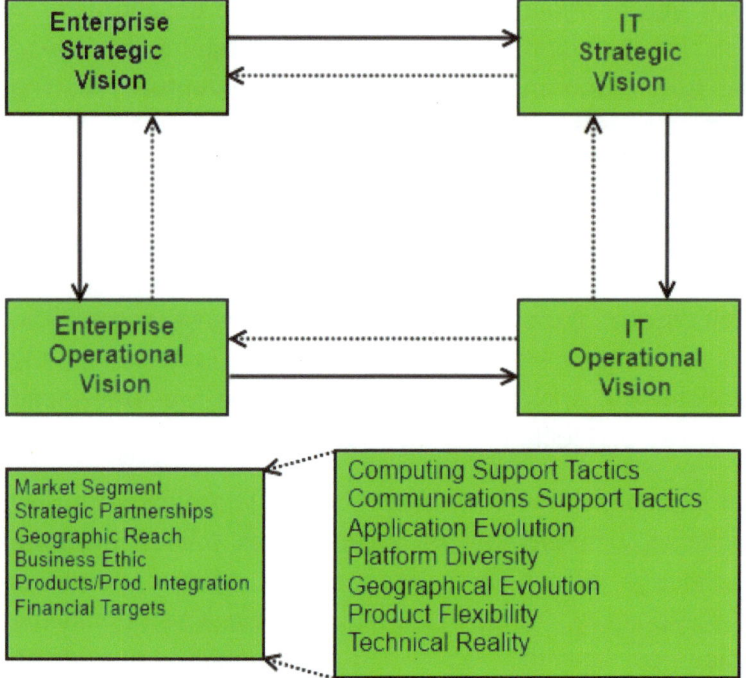

The Vision development process outlined here results in a feedback system that works cross functionally as well as up and down the organization.

This feedback process could go on indefinitely but the executive team will have to be wise enough to call a halt when the process reaches the point of diminishing returns. In the end, the core element and the independent variable remains the Enterprise Strategic Vision. Everything else is a dependent variable supporting and enhancing the Enterprise Strategic Vision.

The Necessity for Integrated Vision

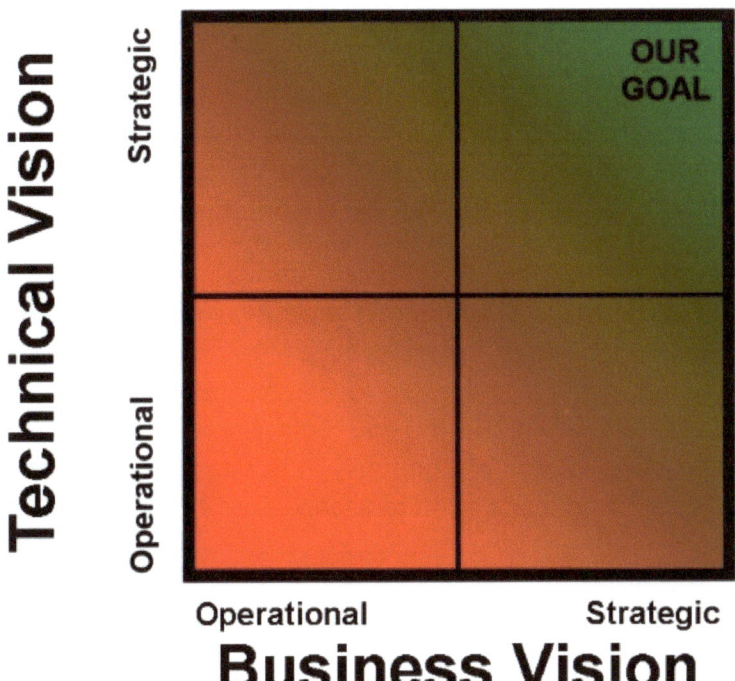

The Danger Area

Whether an enterprise is new or well-established there is the ever-present danger of settling into a solely operational mindset while neglecting strategic initiatives. This can easily happen when the business is functioning smoothly, the customers are happy, margins are good and the employees are satisfied. Perhaps the Strategic Vision that was set in place in a previous planning cycle has been achieved and become the operational routine.

This is dangerous for many reasons. An organization in this mindset can become so comfortable that strategic planning can be the last thing they want to do. All is well with the world, so why increase work by adding long-term planning on top of day-to-day activities? The danger in this is that you can bet that at least one of your competitors has stayed hungry and that strategic planning is a routine, recurring task for them. That competitor will, to the surprise of the complacent and well-run company, zip so far into the lead that the enterprise that gave up routine strategic planning may find itself on the path to failure.

This can create another problem as well. The competitor who is planning for the future is probably also working hard to make sure that the plan includes improved ways to increase satisfaction. So, by not planning effectively for the future you might be behind your competition in at least two different ways. You will be firmly in the danger area and it will take tremendous effort to escape from it.

The danger area we are talking about here is the red zone on the previous chart. The Business Vision for the complacent company described above has become short range or perhaps almost nonexistent. As we have discussed before, Business Vision is the independent variable which Technical Vision is dependent upon. If the Business Vision is operationally, or short-term, focused then the Technical Vision has little to no chance of being Strategic. Unless a senior executive breaks the enterprise out of this red zone the organization will eventually face oblivion.

Danger Area Breakout

Breakout can only be in one direction. Someone at a senior level must cause the team, or some assigned portion of the team, to take on the task of creating a Business Strategic Vision. Organizations that have made

Strategic Planning part of their recurring routine will have established a Strategic Planning mindset, so the planning process will have become standard for them. However, the organization that has been stuck in the red zone for a time may have lost its Strategic Vision skill. This skill may have been lost through simple lack of use, or the people with a knack for Strategic Planning may have lost interest in the day-to-day routine and left for a more progressive enterprise. Whatever the reason, such an organization will find that breakout from the red zone is difficult.

In fact, there is only one logical direction for this breakout. As described in previous pages, Technical Strategic Vision is best built on Business Strategic Vision. An excellent IT staff can estimate a Business Strategic Vision where none has been supplied by the core enterprise team. However, it will only be an estimate and probably based more on technology than focused on the core business. An IT team in an organization which is already in trouble from being stuck in the danger zone is already functioning with a great handicap. That IT team's Strategic Planning task will be made almost impossible if they are also trying to estimate a Strategic Business Vision. So, the best action plan for breakout from the red zone is for the C-level executives to assign a group from the core enterprise to develop a Strategic Business Vision.

Hopefully, your executive level business leaders will not be habitually focused on day-to-day or operational issues. For your business to be successful they must have a Strategic Vision. That seems obvious, but we have probably all, in both the private and public sector, encountered a surprising number of executives who operate from a task list instead of a long term Vision.

I told you about the chief executive who kept a list of tasks that he reported on to our board each month. He never produced a Strategic Vision. We were stuck in the red zone with no hope of moving to Our Goal in the top right corner of our chart.

The point is that moving along the Escape Path takes specialized talent that might have to be brought in from the outside if the organization has lost the necessary skills by staying in the Danger Zone too long. It should also be noted that a company that has developed a Strategic Business Vision is still not fully in the green zone. That requires the integration of Technical and Business Strategic Vision.

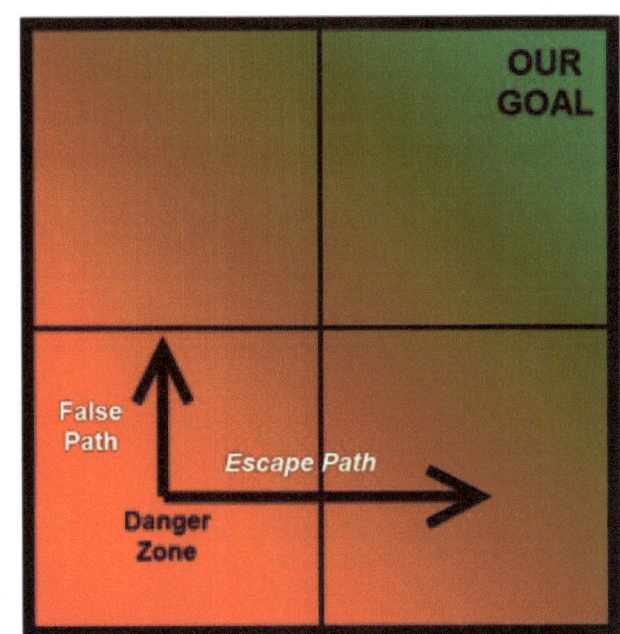

Even the best IT Strategic Planning team will only have partial success at putting together a Strategic Technical Vision with only red zone data. Such an effort will be helpful to the organization but will have little chance of being truly Strategic. However, even an independent effort on the part of the IT team will at least move the enterprise, or at least the IT team, a bit toward a greener area. Even so, any path out of the Danger Zone that does not involve development of a Strategic Business Vision is doomed to limited success and will be Operational more than Strategic.

This independent initiative by an excellent IT team might be called the False Path. Like any false path it is truly a dead end. Any move along the False Path can't break out of the red Operational region because it almost certainly contains more elements of IT Vision than Business Vision. This isn't to say that such an effort should never be made in the absence of a Strategic Business Vision. A False Path initiative will still be helpful because it will move technology forward. The problem, as will be described later in this book, is that the False Path is almost certainly not efficient in developing a Tactical Path to a Strategic Target that achieves Our Goal. The False Path will be repeated over and over creating excess cost and loss of efficiency for the enterprise.

Breakout and Feedback

If the enterprise takes the proper Escape Path from the Danger Zone by writing a first draft of the Strategic Business Vision, then the IT team has the opportunity to construct a legitimate draft Strategic Technical Vision. It is legitimate and in the Strategic Zone because it is based on a real, not guessed-at, Business Vision. This gives the IT team the opportunity to evaluate emerging technologies against forecasted business needs, not just technical evolution in general. Technologies can be selected that not only support the enterprise but can perhaps help the enterprise to move a bit more deeply into the green zone. The feedback provided by the IT team can result in a more aggressive and more competitive Business Vision leading to improved customer satisfaction, lower cost of operation, better product quality and improved margins. This sets a new platform which the IT people can use to further enhance their Strategic Vision statement which will be an improved guide for their selection of even more appropriate technologies. This is an opportunity that would never be available if the False Path was taken. The process continues until the goal is reached.

Mission is the reason an enterprise exists and is very stable and very clear over time. Over a very long period Mission will, of necessity, evolve. For example, a company dedicated to making buggy whips or camera film would not have the same market today that it had in the past.

The Vision of how to accomplish a Mission changes more frequently because of a number of factors. As indicated earlier in this book, an information technology executive will probably find that the IT Vision and technology evolve at a much faster pace than most organizations as a whole. That is true for almost all of the information technologists across a broad range of organizations. The exception is those IT people who are working for an information technology enterprise or other organization evolving at a similar pace.

This is both a challenge and an opportunity. If IT is evolving faster than the enterprise then the IT team has a tougher challenge to keep up with change. However, it is just this rapid change that provides the heads-up team with a great opportunity to provide new tools to help the enterprise excel in its chosen competitive marketplace. If both the enterprise in general, and the IT team specifically, work together in a recurring planning cycle there is great opportunity to function with greater efficiency and effectiveness and perhaps dominate their market.

Planning Frequency

If there is too much time between Vision planning cycles then the organization will have a tendency to drift back toward the Danger Zone. Strategic Goals change over time as a result of market forces and, in keeping with the focus of this book, the rapid evolution of information technology. A reasonable planning interval has to be chosen for Vision review and development. Experience shows this to be about three to five years. More frequent visits to Vision development are generally wasted effort or indicate a Vision that was more Operational than Strategic. Less

43

frequent visits create a greater opportunity to drift back to the danger zone by some amount.

The Optimum Vision Development Path

Our goal in this book is to focus on using the best possible process to develop a Technology Strategic Vision that supports and enhances the Enterprise Strategic Vision.

It is clear from the preceding discussion that there is some optimum interval and method for periodically updating the Strategic Vision. While experience shows that the interval is three to five years for most businesses, there are exceptions. And there is always opportunity for some tweaking of the Strategic Vision between formal planning cycles. It doesn't matter if the planning team is developing the initial Vision for an enterprise, or working on the escape from the Red Zone, or simply updating a current Vision, the process is the same. This process is one that, when applied in the proper manner in appropriate intervals, creates a near linear evolutionary path for the enterprise.

In the first year that you write your Vision Statement and supporting steps you might establish a Strategic Vision that is three to five years away. The second statement might be written a similar number of years after the first plan year. The business will have changed a bit and so will technology. So the Goal might move to the right or left a bit (imagine being at a rifle range) and it will also move a couple of years further into the future (or downrange).

This will happen repeatedly over the years. Let's call the path to the Goal the Tactical Path. If we plan well, the Tactical Path will be smooth and evolutionary. Corrections to the Tactical Path will be minimized and so will costs. Conversely, if we don't plan well, or with sufficient frequency, the Tactical Path will be riddled with discontinuities and significant technology changes that will increase costs and have a negative effect on enterprise competitiveness.

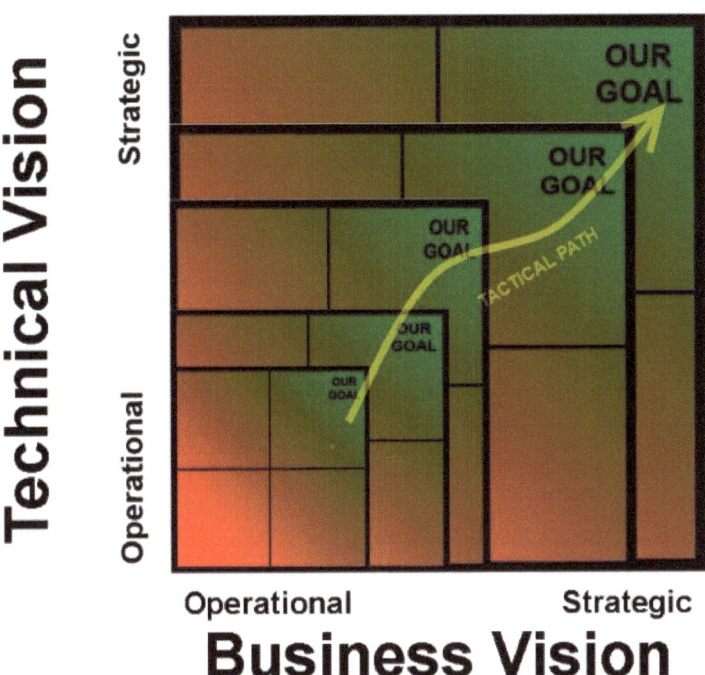

If we plan with sufficient frequency and thoroughness the Tactical Path will have minimal mid-course corrections. A Tactical Path that weaves wildly back and forth is one that reflects a poorly defined Vision or a Vision that is revisited without sufficient frequency. This is not to say that the Vision has to be redefined in each planning cycle. In fact, an excellent Vision may remain the same over multiple planning cycles and only be revised when it has been almost fully attained. Any new Strategic Vision will, in all probability, build on the successes of the previous Strategic Visions.

The Distance To Our Goal

Three to five years has been repeatedly suggested in this book as a proper planning interval. This is based on several decades of experience in a variety of enterprises. In my experience in military, Fortune 150 corporations, regional government and private consulting for a wide variety of enterprises three to five years is pretty much the sweet spot. Anything beyond five years and an enterprise has the tendency to slip back toward the Red Zone. This seems to be true no matter what the business or government agency. A planning interval of less than three years may be necessary due to unexpected changes in economy, your business sector or technology breakthroughs. In an enterprise where the primary product is technology something closer to three years, sometimes even less, may be the standard.

With some exceptions, planning intervals of less than three years will generally mean the previous plan was more operational than strategic.

Even into the early 1980s it was possible to forecast technology in support of an enterprise ten years into the future. The increasingly rapid evolution of technology since then has made this impossible with the rare exception of very long range plans for new developments. An example might be the research being done on fusion power which is a decades-long process.

Placing the Strategic Target too far in the future can have a very misleading result

Scientists from the RAND Corporation have created this model to illustrate how a "home computer" could look like in the year 2004. However the needed technology will not be economically feasible for the average home. Also the scientists readily admit that the computer will require not yet invented technology to actually work, but 50 years from now scientific progress is expected to solve these problems. With teletype interface and the Fortran language, the computer will be easy to use.

The above picture was published by the Rand Corporation in the mid-twentieth century. It illustrates the hazards of choosing the technology component of a Strategic Vision. It is certainly not meant to be critical of the Rand Corporation which in the early 1950's projected what a home computer would look like in 2004.

The picture makes it clear that Rand Corporation missed the mark badly. But they never had a chance of hitting the mark! The pace of technology in the early 1950s and before was dramatically slower than it was in the last half of the 20th century. Few, if any, of the scientists foresaw the invention of the transistor, the large scale integrated circuit, the inventions that would come out of the space program, or the effect of Moore's Law and other technology "laws".

Some of these helpful "laws" concerning technology evolution that have evolved over the last few decades give us what seems to be a pretty accurate picture of the direction in which technology is going. If these "laws" had been understood at the time that the Rand Corporation made their home computer prediction it would be interesting to see how close to the mark they might have come.

Many information technologists are familiar with Gordon Moore's Law which was proposed in 1965 and which has had a few different formulations and interpretations over time. In summary, the thrust of this law is that computing chips will see a doubling of real processor power every two years on average. In retrospect, this has been pretty accurate. One measure reveals that there has been a doubling of real computing power every 2.3 years, on average, since the birth of modern computing.

Then there are the other "laws" which affect our view of the IT future.

- Kryder's Law states that density of information on digital storage devices has been doubling every 23 months, on average, since 1956.
- George Gilder originally observed in the 1980's that "bandwidth grows at least three times faster than computer power".
- Martin Cooper noted that the spectrum efficiency of radio communication (both voice and data) has doubled every two and a half years, over 104 years, since radio waves were first used for communication.
- Nishimura's Law can be interpreted to say that video display size available at the same cost doubles every 3.6 years.

I could go on, but part of the point is made that the Technical Vision is dependent upon some very fast changing basic elements of our environment.

Another part of the point is that inventions along each of these fast moving development tracks work to influence each other and combine to allow engineers to create new technology applications at prodigious rates. Happily, we now have these proven "laws" and can anticipate the general direction and rate of change of technology. For example,

- if you knew from the Business Vision and Infrastructure that your business needed very high bandwidth between two locations, and
- you knew that the bandwidth requirements would double every two years, and
- you expected a long lifetime for your system, and
- it was already operating at 10s of Megabits per second,

then you might choose a wired (fiber) transmission system over a wireless (microwave) transmission system.

Your technology response depends on both the needs of the business as well as the anticipated evolution of the types of technologies available.

Given this and other information we might predict that today's $3,000 desktop computer will in the year 2020 cost $1/500^{th}$ of what it costs today, weigh no more than a cell phone, and be able to communicate wirelessly from a many locations at a rate adequate to handle **today's** applications. Since application bandwidth requirements are evolving faster than wide area wireless bandwidth transmission capability there will probably still be a need to reach the rest of the world through a wired backbone accessed by regional or local wireless access points.

However, since we have learned from experience than any prediction beyond three to five years can be expected to be challenging even with the help of the technology "laws". It would be fun in 2020 to look back at this technology law-based forecast to see whether or not we made a better prediction than the Rand Corporation. While most, or perhaps all, of these laws mentioned above may be accurate there will be other new elements of technology that may make any prediction out to 2020 laughingly inaccurate.

However, this entire discussion in this section devoted to technology doesn't address the implications of technology for the business each of these technologies is being applied to. It is a great, and fun, discussion. But, it is a useless discussion from a business perspective because it didn't refer to concrete business needs.

Business First; Technology Second

I'll repeat here something that many IT people might consider heresy. **Technology without business application has no value!**

In fact, I'll add to the heresy by saying **a CIO does not have to be a technologist**. A CIO must be an excellent business person and an excellent leader. The best CIO I ever worked for had a degree in law and experience in managing paper manufacturing operations. By his own admission, he was the "poster child" for a person in the highest information technology executive position without any IT background at all. He was also the best leader I ever worked for anywhere.

A great deal of his success was based on the fact that he was (he has since retired) a highly intelligent and gifted leader of people. He also understood business and knew that the only reason technology existed in his enterprise was to make the core business better at producing product and customer satisfaction than the competition. To do this it was much more important to understand the business than it was to understand information technology. After all, as CIO he had a highly gifted team of IT people with a wide range of experience and aptitudes whose jobs were to be experts at their piece of technology.

This CIO understood the principles of the charts shown in this book. In fact, he strongly influenced major parts of this book.

The principles of Vision development are pretty straightforward.

- Understand the business first.
- Make sure the IT team understands that their sole purpose within their enterprise is to serve the core business.
- Lead the IT team toward developing IT infrastructure and applications that exist only to serve the business.
- Integrate the IT team at all levels with the business so that each IT team leader fully understands how to make sure that their team serves the business.
- Encourage the IT team to seek ways to make the core enterprise great by using IT applications to enhance every element of the business.
- Develop so much confidence by the core enterprise in the IT team that they are willing to put them in front of external customers so the IT experts can understand what it takes to excite and delight those customers.

Leadership Before Technology

I've been fortunate to work in a variety of leadership positions in demanding environments. During that time I've participated in hiring a lot of people into technology leadership positions. Sometimes I was the hiring executive. Other times, I was a participant or consultant.

As consultant I have witnessed a repeated mistake in hiring at the director level and above. This mistake has often resulted in failure to hire excellent talent. The mistake is the decision to choose someone with a strong technology background instead of a person with excellent leadership skill or credentials. The reason I have heard for rejecting an experienced, professional IT leader is often stated as "He/She isn't technical enough."

The result I've observed in these enterprises is a disconnect between highly technical people in leadership positions who don't know how to lead and followers who are working much too hard in an inefficient environment.

When I was younger I was in one of those organizations. We worked very hard and we worked long hours. We had many technology successes and we certainly learned a tremendous amount about the technology we were implementing. But, as I later realized, the enterprise and the customer were not receiving as much return for this hard work as they deserved. We were like an inefficient incandescent light bulb; ten percent light and ninety percent heat.

For reasons somewhat related to this situation I decided to leave this organization. Through some good fortune I was invited to move to the corporate headquarters which had a much different product focus. It also had a much more enlightened leadership approach. The group that I moved to placed leadership skill first and technology skill second. In fact, it was this group that was led by the CIO mentioned earlier who had no technology background at all. That doesn't mean that technologists weren't put in leadership positions. It means that leadership in executive positions was valued far above technology skill. If both were present in the same individual then that was a happy plus.

What a pleasant change! What I quickly discovered was an environment in which a person could get vastly more work done with a more reasonable amount of effort. We were customer focused rather than technology focused. It was a stark difference.

Leadership and workers mutually respected each other and fully knew what was expected. There was a relaxed confidence coupled with a desire to produce excellence. Pride rather than exasperation was evident everywhere.

Leadership was the key. It started at the top and it trickled through the entire IT management structure. The IT leadership was respected by the CEO, the board and the division presidents. This respect by the core enterprise for the IT leadership was in great part what opened the doors necessary for the IT team to be enthusiastically welcomed into the divisions.

The point is this. The core enterprise leaders regarded the IT team as business leaders who happened to have a strong background in information technology. This enabled an open dialogue between people who had a mutual respect for each other's skills. This dialogue, through the iterative process described in this book led to an integrated business and technology Vision that enabled the organization to succeed, and even excel, in a very competitive business.

In this corporation with talented leaders it was standard practice to periodically create and implement Visions for the future that integrated the information technology team into the core enterprise to help the enterprise achieve greatness.

Within the IT team, the people who studied and recommended the Strategic IT Vision were senior manager and director level people with strong business and leadership skills and a solid background in technology. Business and leadership skills were the primary requirement. Technology skill was necessary but a secondary priority. The Vision team could easily draw on the detailed technical experience of the teams they led. However, they couldn't effectively apply that technical experience unless they had an aptitude for business and a strong ability to apply technology to business.

That was a key ingredient of skill for that Vision team. They knew **that technology without business application has no value!** Every step of the Vision development process was made with that major premise in mind. That mindset started with the CIO and trickled down through middle and first line management to every front line worker on the information technology team. The entire orientation of the team was toward the business. The Strategic IT Vision was based on the Strategic Business Vision because that was the culture of the organization as cultivated by the CIO.

This was in great contrast to the organization I had left where people in executive technology positions were chosen first for their technology skill and secondarily for their leadership skill. In fact, many of the executives in these leadership positions had little business savvy or people skills. Strategic Technical Visions were weak because the IT executives lacked the skill and experience necessary to understand the core business they were supporting. Their employees worked hard and put in long hours but they were simply implementing new technologies with limited understanding of the business reasons for the work they were doing.

The difference between the two organizations was stark. The employees in the IT organization with strong business leadership skills and a business oriented Strategic Vision were much more efficient and effective. Employees worked a normal work week rather than massive amounts of overtime and got more done! The core enterprise received great business benefit from the application of technology. From another perspective it was also interesting to note that technology team employee morale was high and turnover was exceedingly low.

The primary lesson in all of this it that **excellent leaders create excellent organizations!** More to the point for this book, excellent IT leaders lead teams that create great Vision statements that enhance the ability of enterprises to excel.

The One Page Guiding Document

When you create a Strategic Vision you are essentially saying to your team "We are going to move from where we are today to this new place that I have described for you". The only reason for an enterprise to move to that new place which we call the Strategic Target is that there are some business Benefits in doing so. Obviously, to reach the Target and achieve those Benefits some Actions have to be taken. All of this brings us to a point that I like to constantly remind people of. It is important to keep things simple because they will get complicated all by themselves! This book reflects strong adherence to that principle.

Keep it simple because it will get complicated all by itself.

Once all the work described in the previous pages of this book has been accomplished a one-page document can be published that summarizes where the chief technology executive is taking the team.

When that one-page document is completed everyone should be given a copy, and it should be placed on the wall as a large poster that is in full view of the entire team.

An **example** of this one-page document is shown on the next page.

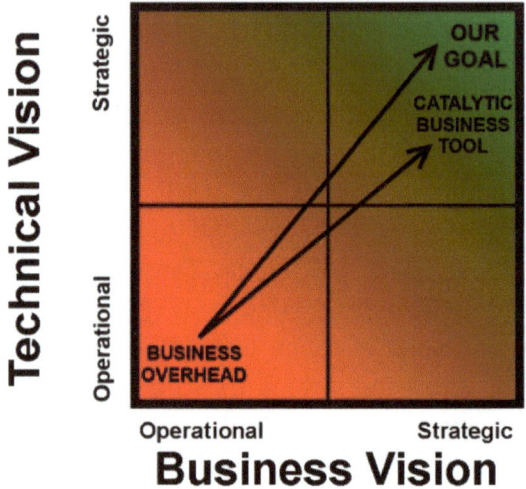

VISION – We will work as a team to move our systems and products:
- From Business Overhead Elements to Catalytic Business Tools
- From Discrete Technical Tools to Sales Acceleration Systems
- From Back Office Support Elements to an Essential Resource that Pervades Every Part of the Enterprise
- Etc.

BENEFITS - Our technical products and services:
- Will provide . . ..
- Will be . . ..
- Etc.

ACTIONS - As a team we will take these Actions to reach this Target and achieve these Benefits:
- Leverage . . ..
- Provide . . ..
- Implement . . ..
- Etc.

This **example** document contains a Vision matrix at the top of the page that describes:

- where the technology team is today on the Vision matrix,
- where it must be on this matrix at the end of the planning horizon,
- the operational service the team provides today, and
- the Strategic service it is expected to provide at the end of the planning horizon.

In the interest of full disclosure, it is important for me to point out that there is more than one type of matrix that can be used at the top of this chart. The matrix type used depends on the method you choose to reach this point in the planning process. Some will choose to reach this point through the development of an IT Strategic Plan. (See my book "How to Write A Great Information Technology Strategic Plan – And Thrill your CEO) and use the matrix in that book. The important thing is to create a one-page document that is the simple motivational description of the goal and how the team is going to reach that goal.

In the example I've used the IT team planners have recognized that information technology within the organization is serving only as an essential overhead revenue consumer. The IT team has the goal to move to being a core business enhancer.

Underneath this matrix is a bullet point description of:

- the Vision,
- the Benefits that accrue for the enterprise, and

- the Actions that must be taken to achieve those Benefits.

This one-page chart tells the IT team what they need to do. It also tells everyone in the enterprise, in clear business terms, how and why the IT team is engaged in the continuing set of enterprise enhancing actions they are taking.

The VISION paragraph combined with the first bullet under VISION can be the Vision Statement sentence. Or it might be the first line in a short Vision Statement paragraph that includes the information in all of the bulleted subparagraphs under VISION.

The core Vision Statement is clear. "We will work as a team to move our systems and products from business overhead elements to catalytic business tools."

The first bullet under VISION defines the primary objective of moving from the main description of where to you are today to where you plan to be at the end of the planning horizon.

The remaining bullets under VISION provide a bit more granularity to describe elements of the current environment that you intend to move away from. You may want to add these to your basic Vision Statement.

This one page chart is not just a definition of where you are leading your team. It is also a description of why you are taking them there. The BENEFITS paragraph and its bullets provide that description. This is the motivational section and helps to keep your team constantly informed about why you are driving toward the Vision

Finally, the one page chart states in bullet point form the ACTIONS you and your team will take to reach your goal. These must be succinct yet definitive.

You should work to make all of this fit on one page. That's tough but it can be done. I have done it several times. It is important that you also put this one-page document on poster boards placed in conspicuous locations. Your team will then always be reminded about the Vision, Benefits and Actions.

Most importantly you, as the team leader, must always be selling the Vision and showing your dedication to that Vision.

If this process is followed, there will have been an extensive amount of background research done by team members to develop the Vision Statement and establish a path to reach the Mission Vision. This one-page document that results from all the planning and research is the simple tool everyone will refer to as the team moves forward.

In my experience, a well-led team, clearly briefed on the Vision Statement and the meaning of this one-page chart, will aggressively and enthusiastically accomplish their mission.

People will meet your expectations.

Remember that the first and most important rule of leadership is that "People will meet your expectations". This one-page leadership document sets the expectation of success. It keeps the Vision in full, clear view in simple terms.

It is clear to all of us that there is a lot of work behind this one-page document. The work involves a close partnership between the CIO's team and the core enterprise planners. The IT team must make a great effort to fully understand the Enterprise Vision and Infrastructure and ensure that an Information Technology Vision and Infrastructure is created that supports and enhances the enterprise objectives. The CIO must be a true leader who excites the team and leads them with enthusiasm to achieve absolute success.

There was also a lot of work that resulted from President Kennedy's simply stated vision of "before this decade is out of landing a man on the moon and returning him safely to earth". He stated an expectation, and the nation met his expectation!

The one-page planning document defined here sets an expectation that is clear and concise and can be easily made visible to all. It provides an achievable Technology Vision that is fully based on, supports, and enhances the Enterprise Vision and Infrastructure established by the enterprise executive team.

About The Author

Tom Ireland was an "Air Force Brat" bouncing around military installations in the U.S. and Far East. He often says that growing up in the military was a great adventure. It was during this adventure that he *spent three of his high school years in Japan.*

Tom received a BS in Electronics Engineering and a minor in mathematics from Oklahoma State University. While there, he published his first paper in "The Oklahoma State Engineer". That paper won first place in a Big Eight Universities competition.

Tom accepted a commission in the United States Air Force. While an Air Force Officer, and later as a Department of Defense civilian, he managed technology program management teams responsible for communications and electronics installations in the Middle East, Wright-Patterson Air Force Base, Ohio and other military installations in the Ohio region. His Plans and Programs duties included disaster preparedness; survival, recovery, reconstitution and war planning.

After leaving the Air Force, Tom went to work for a technology division of the Mead Corporation responsible for designing and operating their networks in support of what was at that time the world's largest on-line, full-text search and retrieval data base. He eventually moved to Mead Corporate Staff where he successfully integrated their network operations and directed the modernization of the corporation's client computing services systems. He participated in a Vision Team which successfully established a long term IT Vision for the corporation.

Tom later went to work for CompuServe/UUNet. Over three years he integrated the operations of their Class A web hosting data centers while the operation was expanding from three to twenty data centers across North America.

Tom is now the Chief Technology Officer for a technology team supporting several cities in southwestern Ohio. He also provides technology support to the regional Tactical Crime Suppression Unit and the Organized Crime Task Force.

He has been featured in a cover article in "Wireless for the Corporate User", and has been published in "Telecommunications Magazine" and "Community Media Review". He is co-author of an AT&T Labs paper presented in Munich, served as the Bell Labs liaison for the Mead Corporation and has been personally highlighted in an AT&T annual report as an example of collaborative customer innovation.

Tom is married with four children, two "inherited" children and ten grandchildren, some "inherited" also. These days he is indulging a life-long desire to do more writing across a number of subjects reflecting his broad range of interests and experience.

www.ingramcontent.com/pod-product-compliance
Lightning Source LLC
Chambersburg PA
CBHW041105180526
45172CB00001B/111